MEDICAL MEDITATION

HOW TO REDUCE PAIN, DECREASE COMPLICATIONS AND RECOVER FASTER FROM SURGERY, DISEASE AND ILLNESS

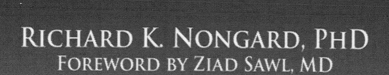

RICHARD K. NONGARD, PhD
FOREWORD BY ZIAD SAWL, MD

For bulk copies and institutional orders, a discount is available on the purchase of 5 or more copies of individual titles for special markets or premium use.

For further details, please contact the publisher:
PeachTree Professional Education, Inc.
7107 S. Yale, Suite 370
Tulsa, OK 74136
(800) 390-9536
www.SubliminalScience.com

Medical Meditation
– How to Reduce Pain, Decrease Complications, and
Recover Faster from Surgery, Illness and Disease

By Richard K. Nongard, PhD

First Printing, January 2010

Printed in the United States of America

ISBN: 978-0-557-25592-4

DISCLAIMER

No book, DVD or CD can diagnose or treat illness. Only a licensed physician, who examines you in person, can diagnose or prescribe appropriate treatments. Users of this book should consult with their physician about the use of exercises in this book, and at no time should they discontinue prescribed or recommended treatment without consulting their personal physician first.

The author of this book is neither a medical doctor nor a specialist in any medical condition. The following is offered as general information only, and as such, may not be applicable to a specific reader and his or her problems or medical conditions. It is clearly not based on actual knowledge about or examination of the reader, and therefore it cannot and should not be relied upon as definitive medical opinion or advice. The reader is instructed to consult his or her own medical doctor before proceeding with any suggestions contained herein, and to act only through his or her own doctor's orders and recommendations. By reading the pages that follow, the reader stipulates and confirms that he or she fully understands this disclaimer and holds harmless the writer. *If this is not fully agreeable to you, the reader, you hereby are admonished to read no further.*

ABOUT THE AUTHOR

Richard K. Nongard, Ph.D., is a meditation instructor, minister, and a practicing psychotherapist who has helped both professionals and individuals to learn the concepts of meditation. His experience includes both traditional mental health hospital and private practice work. He holds degrees in ministry, counseling and religion. His expertise comes from working on physician-led treatment teams as a psychotherapist or as a case manager in mental health settings.

As a lifelong student of comparative religions, his understanding of meditative traditions and the role of cultural healers has played a significant role in his synthesis of this material. In addition to teaching patients the techniques of medical meditation, he has worked closely with family and friends to help them recover from life-threatening illnesses, and has used the materials presented in this book to promote health following his own surgery.

CONTENTS

FOREWORD

Surgery. Illness. Pain. Disease. These words terrify the people who suffer from them. And worse, beyond the direct suffering they cause, these situations can also lead to anxiety, loss of sleep, lack of concentration, and even loss of the ability to focus on productive, meaningful things in your life.

However, my good friend and colleague, Richard Nongard, believes it does not have to be that way, that you can use your own mind and your body's own healing powers to fight back against the negative effects of pain and illness. By using a simple system of meditation exercises, this book teaches you how to minimize pain and anxiety, reduce the stress experienced prior to surgery, and even speed up healing and enhance recovery.

As an anesthesiologist, I have direct experience with stress related to surgeries. Anxiety prior to surgery, even before "simple" procedures, is ubiquitous. Many patients try to hide their fear with a facade of courage and humor, while others simply fall apart. As medical doctors, we use sedatives and medications to ease their fears - but perhaps there is a better, more natural effective way...?

Through meditation, you can teach your own mind to work for you, rather than against you. I am not suggesting that there is anything magical about meditation; it is not laying on hands or curing leprosy with a mantra. It is, however a proven effective weapon you can use to help protect and enhance your health.

And within this text, you will learn how to naturally harness the power of meditation to enhance your comfort, recovery and overall sense of well-being.

Best wishes,

~ *Ziad Sawi, MD*
Diplomate, American Board of Anesthesiology

INTRODUCTION

I have been teaching medical meditation techniques to clients for many years, and have spent many years instructing licensed mental health professionals in how to teach patients the techniques of meditation and self-hypnosis that bring pain relief, promote physical healing and generate a sense of well-being during difficult times. Through these experiences I am well-qualified to write a book on the subject of medical meditation, but I had put off writing this book for several years. The reason I procrastinated was not laziness, busyness or a lack of motivation. I wanted to write this book not just as a professional educator and therapist, but also from the perspective of a patient who has used medical meditation both before and after surgery.

Although I have been blessed with overall good health and my personal meditation experiences have focused on developing a wide variety of meditation skills, the last six years have made this topic more relevant to me from a personal perspective and I think this makes the writing of this book at the present time even more valuable. The skills I am teaching in this book have come not only from my collective academic studies or personal practices, but my own applications to the management of chronic arthritic pain, culminating in a rather complex surgery that required a lengthy recovery. While preparing this material, I also shared my experiences and techniques with my lifelong friend and office manager, who has gone through multiple surgeries and medical procedures for the treatment of breast cancer. Both of us have used these ideas during our surgical experiences and, like the many patients I have taught in professional settings, have found the application of these principles to be life-changing.

This book is merely a guidebook: it is short and teaches specific skills. It is designed to be used by a person with no experience in meditation, or perhaps even by people with little interest in meditation who are instead motivated only by the benefits that research has shown to come to those who practice meditation as part of their surgical preparation and recovery process. It is my hope that you will find relief during the uncertain and sometimes painful waiting periods that precede surgery, and a speedy recovery free from complications following surgery. It is also my experience that these techniques are transformational in every aspect of life, and that the improvements you find during this difficult time of merely trying to manage your health will transcend into your personal relationships, your general outlook on life and your emotional well-being and give you a sense of control in any of life's scenarios.

Enjoy!

~ *Richard K. Nongard, PhD*
Meditation Instructor

CHAPTER 1:
WHY MEDICAL MEDITATION

The reason meditation is taught to medical patients is simple: it is a highly effective tool for managing pain, reducing anxiety, helping you to control your body, healing at a cellular level, reducing recovery time and reducing dependence on medication. The benefits of medical meditation are well-documented. It is not just a theory or speculation that learning the techniques of medical meditation or self-hypnosis is extremely valuable to patients. In fact, the research is so compelling that at times I wonder why every surgical patient is not enrolled in a class or given a CD teaching them the methods of medical meditation.

> Meditation decreases oxygen consumption, heart rate, respiratory rate and blood pressure, and increases the intensity of alpha, theta and delta brain waves — the opposite of the physiological changes that occur during [stress].
>
> *Herbert Benson, M.D., Harvard Medical School, author of The Relaxation Response*

John Kabat-Zinn, Ph.D., an associate professor of medicine at the University of Massachusetts

Medical Center in Worcester, has made meditation the centerpiece of the center's Stress Reduction Clinic. "It doesn't seem to matter what type of medical condition brings people to the Stress Reduction Clinic," says Kabat-Zinn. "Over the eight-week program, they usually report a reduction in symptoms."

IDEA Health & Fitness Source, September 2000

Meditation also reduces the impact of several peculiarly Western diseases. Studies have shown that meditation can reduce hardening of the arteries, especially in African Americans with high blood pressure. People suffering from anxiety disorders also appreciate the lowered stress, reduced blood pressure and slowed heart rates that are associated with meditation. Similarly, there is growing evidence that a meditation program can have a positive, sustained effect on chronic pain and mood, including depression and anxiety. In an even more dramatic example, initial research has suggested that meditation combined with dietary changes may slow tumor progression in prostate cancer patients.

Memhet Oz, M.D., Columbia Presbyterian Heart Institute, Time Magazine Jan 20, 2003

Surgery patients who learned simple relaxation and meditation techniques stayed in the hospital an average of 1.5 days fewer than those in a control group. Results include faster recovery from surgery, fewer complications and reduced post-surgical pains. Findings were consistent in 191 independent studies.

University of Wisconsin School of Nursing, Patient
Education and Counseling 1992

Did you read that last line? "Findings were consistent in 191 independent studies" on the surgical benefits of medical meditation alone! Patients experience faster recovery, fewer complications and reduced pain.

Chances are pretty good that someone gave you this book or recommended this book, and up to this point you have had very little interest in meditation, perhaps even seeing it as some Hollywood manifestation of itself, characterized by people sitting in funny positions and saying strange words in a dark candlelit room. But when people read that "findings were consistent in 191 independent studies" that meditation brings the specific benefits you are seeking, it suddenly makes the whole subject much more appealing.

A lot of misconceptions surround meditation and its sister, self-hypnosis. In fact, for the purposes of this book I am going to use the terms interchangeably. Self-hypnosis uses the methods of meditation to help a person internalize the ideals of health, wellness and recovery, and in many ways it is the same process. Self-hypnosis and meditation are not 'surrendering your will to a mysterious force,' but rather tools to help you recognize the body's innate ability to provide healing from within. And although deep levels of relaxation do occur in medical meditation and self-hypnosis, a person can never become "stuck" in a trance state because trance is a natural healing function of the body. Every night we enter sleep through a trance process, and many times we naturally recharge our mental powers with momentary mental breaks that are trance states. Trance is not a state of mindless unawareness such as Hollywood portrays but is the body's own state of natural healing. Although not quite the same as the natural trance states that occur on a day to day basis to all of us, victims of trauma are often placed for days or even weeks into artificial comas in many intensive care medical settings specifically to enhance the body's healing process.

There are other misconceptions about meditation that hold people back from deriving the full benefits of this natural ability. The myths include a belief that meditation is always inherently religious and is usually associated with religions of Eastern origin. Although a historical study of meditation does reveal that the origins of many of these techniques may have come from ancient tradition, this is because in the past, spiritual healers were the only physicians that existed since modern medicine did not. As a result, ancient traditions studied the body and its wellness as much as the mind or the spirit. But it is important to realize that these techniques are in and of themselves just techniques, not "secrets of the gods" revealed only to those on a specific pathway. The benefits of therapeutic breath work help to promote healing for the atheist, the Catholic, the Baptist, the Jew and the Muslim – and for members of the Hindu, Taoist and Buddhist traditions that may have popularized these techniques.

There is nothing inherently religious in medical meditation, although those who practice a religious faith will find that the practices of prayer, ritual and meditation that they are familiar with may enhance their experience in learning and applying the principles of medical meditation.

Chapter 2: Defining Medical Meditation

I want to spend a few more pages articulating what medical meditation is, and what it is not. Then we will quickly move on to exercises that you can begin to use today, and that will bring you the promised benefits of freedom and health.

What Medical Meditation is Not

Medical meditation is not a quick fix. It is a tool or lifestyle choice that has to be learned and practiced. Although it is true that the learning comes quickly and the benefits can be immediate, continuing to add to your experiences and incorporating the practice sessions into automatic responses is key.

Medical meditation is not toxic. I have never been opposed to the use of medications to enhance wellness and recovery. However, most medications, even natural herbs and cures, have both risks and side effects. Meditation is from within: trance states or relaxation responses are something that are innately part of who we are as people. As a result, the use of meditation can decrease the dependency on medications, including pain medications, the need for anesthesia in some instances and the use of psychotropic medications or even alcohol and other drugs commonly used to self-medicate during periods of medical difficulty.

It is also important to note that meditation is not a technique requiring superhuman feats, movements, postures or unusual positions. Meditation is often associated with the posture of the Swami, the lotus position, or movements from yoga. While these poses, postures and movements have purposes and benefits, there isn't anybody who cannot practice meditation due to physical limitations. My last surgery was on the foot. Prior to surgery, and certainly for many months post-surgery, it was impossible to sit on the floor or to hold a lotus position. I meditated while sitting in a chair. In fact, most hospitals who teach medical meditation instruct patients to meditate sitting in a chair or lying on a bed. There is not a right way or a wrong way to meditate, and working within your limitations is easily accomplished, without diminishing the effectiveness of meditation.

Medical meditation is also not for those who do not want to change. Over my many years in a variety of medical settings, I have worked with patients who chose not to comply with medical procedures, essential medication or the advice of staff and doctors. For some reason, some people choose to stay sick, and in many cases get worse. Every physician can tell you stories about patients who could have made recovery from illness but who for some reason chose to stay ill. This is not to say that all clients can always recover fully, or that all patients who remain ill do so because they choose to, but rather to point out the odd reality that at some level some people meet important needs through their illness.

Medical meditation is transformational, and if you want to get well, change will occur. For some people this change is quick and profound, for others it is slower or less profound, but every patient who applies the principles of medical meditation will experience improvement on some level. It is not for those who wish to stay stuck.

WHAT MEDICAL MEDITATION IS

Medical meditation is a process that uses the body's natural resources on a cellular level. It promotes the body's use of oxygen and blood flow and increases immune system response. At its heart it is a physical activity rather than simply a mental process. Changes in heart rate, respiration, blood pressure and many other physical responses can be measured during meditative states.

Medical meditation is also a mental process. It is a discipline that is learned, and then incorporated into daily life. Joseph Goldstein, in his instructional CD series called *Abiding in Mindfulness*, relates a story of a student wondering if meditation is a waste of time. The teacher answers that meditation is like sharpening an ax. When one sharpens an ax they are not actually chopping down a tree, but when they go out later to chop down a tree it is a lot easier if the ax is sharp. Medical meditation is like sharpening an ax. Although benefits at the time of meditation may be present, when confronted by the unexpected complications of medical problems, having as part of your mental coping strategies and repertoire the tools that are internalized during meditation can make all the difference in the world. During my last surgical procedure, I went into the operating room expecting one specific surgery with a fairly short recovery time, but emerged from my anesthesia having undergone a different operation that immobilized my foot and prevented me from walking for over eight weeks. These situations are not uncommon. Having a sharp ax can prevent depression, fear, pain and deep feelings of insecurity, dependency and even despair.

Medical meditation is something that is easy to learn. Although practicing, refining and experiencing the methods offered in this text may take some time, in our very first exercise you will be able to do something that is helpful to you. The experience will help you and you will be able to replicate this benefit at anytime in your life that it is useful to you. There are no techniques in this book that require fancy positions or poses, no techniques that are difficult to master and no techniques that require anything more than a little effort, a few minutes of your time and the desire to see improvement in both the short-term and long-term.

Medical meditation is a true solution. The research and the results are in. Medical meditation is not just a novelty or an ancillary intervention, but a true solution to many medical complications. The following is a list of specific health benefits from the Mayo Clinic website.

Research suggests that meditation may help such conditions as:

- Allergies
- Anxiety disorders
- Asthma
- Binge eating
- Cancer
- Depression
- Fatigue
- Heart disease
- High blood pressure
- Pain
- Sleep problems
- Substance abuse

Additional research studies show medical meditation having benefits in regards to:

- Anti-aging
- Fibromyalgia
- Surgical preparation
- Diabetes
- Immune disorders
- Wound healing
- Bone fusion
- COPD and Emphysema

Medical meditation is also a spiritual activity. Not from the perspective of any religion or dogma, or even from the perspective of mystical experiences, but rather from the perspective that meeting our deepest needs is a pathway to spiritual fulfillment. Think about your medical conditions for a moment. How have they limited you? How have they diminished your joy for life or brought about a sense of uncertainty in your life and in the life of your family members? Our deepest needs, which are spiritual if you will, are security, significance, freedom and a sense of belonging. By practicing meditation, healing is promoted and wellness is the result. A state of wellness not only produces a decrease in pain but an increase in functionality, and this in turn gives a newfound sense of freedom.

A sense of freedom produces significance, once again allowing you to see yourself as having value and being able to contribute to the world around you. The result is a sense of belonging and security. In the book Alcoholics Anonymous, those who practice the principles of the 12-step program are told that they will receive what the books calls "The Promises."

The Promises are:

- *We are going to know a new freedom and a new happiness.*
- *We will not regret the past nor wish to shut the door on it.*
- *We will comprehend the word serenity and we will know peace.*
- *No matter how far down the scale we have gone, we will see how our experience can benefit others.*
- *That feeling of uselessness and self-pity will disappear.*
- *We will lose interest in selfish things and gain interest in our fellows.*
- *Self-seeking will slip away.*

- *Our whole attitude and outlook upon life will change.*
- *Fear of people and of economic insecurity will leave us.*
- *We will intuitively know how to handle situations which used to baffle us.*
- *We will suddenly realize that God is doing for us what we could not do for ourselves.*

Are these extravagant promises? We think not.
They are being fulfilled among us -
sometimes quickly, sometimes slowly.
They will always materialize if we work for them.

I think that those who practice the principles of medical meditation will find that they too will experience these same promises, and although not religious, these are certainly spiritual experiences for those who have been trapped by medical illness.

Chapter 3:
How to Do Medical
Meditation

In this book, I am going to teach you a series of exercises all meditative in nature, and all designed to accomplish specific goals related to your health. It is important to realize that these methods are not an alternative to treatments that your doctor has prescribed, but rather are in addition to treatments that a physician has recommended. You are strongly advised to discuss your incorporation of medical meditation into your treatment with your physician, and even if you begin to experience profound change and recovery, continue all treatments that are advised by your physician until he or she tells you to discontinue such treatments.

Posture

The starting point is finding a place and a posture for meditation. You should find a comfortable place free from distraction. You may sit in a chair or lie on a bed. What is most important is not the place you choose but your comfort and that you make a conscious decision to take a few moments for yourself to practice the techniques in this book. It is entirely possible to do these exercises at your desk, even with people and work in the background, and I have practiced these techniques many times while waiting for a flight. In the long run, meditation actually becomes a lifestyle, with one practicing the principles of meditation all of the time. But for now, making a decision to practice and finding a quiet place where you can sit comfortably or lie down is beneficial.

When I instruct people, it is almost always in a classroom environment or one-on-one in my office. I almost always teach people to sit in a chair with their feet on the floor, and to avoid crossing the feet or arms. I prefer a new learner to rest the hands on the thighs, and to sit upright on the edge of the chair without leaning into the back of the chair. This is a comfortable position for most, even those with physical limitations, and the erect spine promotes awareness of the body rather than a laziness that can accompany slouching.

Although I rarely mediate lying down, after surgery I was required to elevate my feet for hours on end and often mediated lying down. Sometimes when I am instructing people, lying down is how they prefer to mediate, even though it is not my preference. Again, there is not a right way or a wrong way, but some breath work exercises are better served with more of a formal posture than a relaxed posture.

BREATHWORK

Breathing is the only thing we do from the moment of birth until the last moment of our life. Everything else that we do we start doing and stop doing at various points in life. Some people start playing a sport at a young age, and retire from that sport when the business of life impedes the ability to play. Other people begin participating in a hobby when it is interesting and stop when funds run out or they get bored. At other times in life we may start doing something and stop, only to resume doing it again. Smokers are notorious for doing this. But breathing is different from everything else we do. We do it from start to finish in life, never taking time off to not breathe.

Since breathing is natural and something we are experts at, it may seem silly to focus on breathing, but it really is the starting point for medical meditation. This is true for two reasons, the first being that proper breathing promotes physical wellness and restoration of the cells. It promotes healing and ensures circulation of both oxygen and blood through every fiber of our body. The second reason is that because breathing is automatic, it makes a good focal point for us in learning mindfulness – a key concept in medical meditation.

Mindfulness has been defined in many ways, but a good definition is simply "paying attention to something in a particular way." Rarely do we focus on our breathing, unless there is something wrong or we are short of breath. In mindfulness meditation, focusing on something as mundane as the breath gives us a practice (and meditation is practice) and the ability to focus on other things.

In meditation, one might focus on the breathing, noticing how air feels as it travels through the nostrils, and the sensation of air entering the lungs. In a meditation, one might also focus on the speed or feeling of exhaling, all tasks we do mindlessly many times each day. What is the value in this attention to breath? It helps us to be present, and present living is essential for the medical patient.

So much of our time in life is spent reliving the past, our mistakes, our actions and our difficulties. Many people find that if they are not ruminating on the past, they are projecting into the future. They project fears of change, worry about the "what-ifs" and how to control the outcomes of people and situations that in many cases they have no real control over. It is the meditator who has mastered a new manner of living, a manner of living based in the present. Ultimately, the only moment in life we have is this moment, and meditation exercises focusing on the breath are a starting point for present living.

Living in the present can end anxiety, allow the body to rest and begin a profound change in perspective. My favorite movie line of all time comes from Oogway, the wise turtle in the movie *Kung Foo Panda*, when he said, "You are too concerned about what was and what will be. There is a saying: Yesterday is history, tomorrow is a mystery, but today is a gift. That is why it is called the 'present'."

LEARNING TO BREATH CORRECTLY

Before you read any further, take a moment to intentionally take a deep breath. Go ahead, do it now. Inhale deeply and exhale.

Years ago, I sat in on a class being taught by a well-known nurse anesthetist who teaches other nurses the skills of pain management. He asked the participants, who were all medical professionals, if they knew how to breathe. Everyone raised their hand to show that they did. He then asked everyone to take a deep breath, and everyone inhaled and collectively exhaled. "Ah! You did it!" he said. "You breathed just like everyone breathes when I ask them to take a deep breath." Everyone looked a little puzzled, and he went on to explain that what most people did was stretch out the back, puff up the chest, tense the neck and inhale, and then exhale by collapsing the chest and slumping. This is what I had done also. He explained that in doing so, we only allowed oxygen to move into the top part of the lungs, and had tensed our muscles and therefore missed the true benefit of a proper breath.

I had never thought about it before, but I, like many people when intentionally taking a deep breath, didn't know how to do it correctly.

He then went on to have us sit in an aligned and upright position in our chairs. He asked us to close our eyes, and imagine that inside of our stomach was a balloon, just an empty balloon, but a balloon like that a kid might have before it is filled with helium. He told us to envision now that this balloon was being slowly filled with air, with us inhaling as the balloon became larger. When it came time to exhale, he had us envision that the air was being let out of the balloon, teaching us to breathe with our belly or diaphragm. What a difference! No tension in the back of the neck, no puffed up chest, and I could practically feel the oxygen reaching the lowest part of the lungs. The difference was night and day.

I work with a lot of people trying to quit smoking in my hypnotherapy practice. I have found this simple breathing technique to be the most valuable technique that I teach them. Even before I begin a formal hypnotic induction, I always teach them this method of deep breathing. In hypnotherapy, I then make the suggestion that they will intentionally take time several times a day to practice this method of deep breathing. For the smoker, oxygen becomes an energizer that replaces nicotine. This deep breathing promotes the restoration of heart and lung functioning, and it is a technique that simply stops withdrawal symptoms during those first 96 hours of being a non-smoker.

But who else can benefit from this simple breathing exercise? Many, such as medical clients who need healing on a cellular level and patients who are managing hypertension, high blood pressure or pain. Breathing promotes living. Patients who want to live life, and live it to the fullest, learn to practice deep breathing as a practice of restoration.

EXERCISE 1:
BREATHING DEEPLY

Take a few moments now to sit on a chair in a quiet place. Allow yourself to be free from distractions for a few moments, turning off the phone or closing the door. Now scan your body, and anywhere you are holding tension let the tension melt away, relaxing the muscles in your body.

Take a few moments to practice deep breathing. Inhale through your nostrils as you envision that balloon being filled with helium, and then exhale by envisioning that air being let out. Repeat this several times.

Congratulations! You are now a meditator. At a basic level that is all there is to it. If you can do exercise one, you can do them all. You have done it!

This is a very basic exercise but also a very important exercise, and one that marks the beginning of true change.

CHAPTER 4:
DEVELOPING
MINDFULNESS

The idea behind mindfulness is twofold. First, it is present living. Mindfulness means paying attention to the present and giving attention in a specific way. What is this way? Nonjudgmentally.

Let's take the problem of pain for a moment. The moment you realize that pain is increasing, you attach meaning to this in a judgmental manner and say things to yourself such as, "I'm not going to be able to work tomorrow," or "I'm so depressed because of this pain." In life, it is almost as if we have been programmed to attach meaning to experiences, but one who truly lives in the present simply sees the present as it is, without fear, without projection and without judgment.

When I emerged from surgery and realized that the procedure that was done was much more complicated than the surgery I anticipated, my natural inclination was to judge my predicament. Still hazy from the anesthesia, I said to my mother, "Oh no! I only took off four weeks from work!" I thought to myself upon returning home, "But I thought I would be able to walk again quickly, and now it will be months!" These thoughts, although natural and understandable, are judgments. They decrease the quality of life, because even if true, they are not something that I can control in the present.

Present living is much better than living in a future projection of "what-ifs." The idea behind mindfulness is not to ignore stark realities, or unpleasant thoughts, but to see them as they are - simply thoughts at this moment.

We all know that our thoughts about the future often do not materialize in they way that we anticipate them. Many people go though life expecting to work hard or even just scrape by, but then experience a windfall and are pleasantly surprised to find that life is filled with rewards that were never expected. Other people expect the good fortune they have always had to follow them throughout life, only to discover health, finances or relationships did not work out the way they planned.

Learning mindfulness is an essential skill in medical meditation. For those facing surgery, it can decrease anxiety or worry, and for those managing pain, it can push aside the realization of long-term complications and let someone simply experience the here and now as it is. It can change pers-pectives on health and wellness and even on illness and death.

Perhaps you have heard about religious monks who meditate for hours and hours each day. Is that what it takes to be truly mindful? Fortunately the answer is no. You can begin right now practicing mindfulness, and although there is a popular (and excellent) book called, "The Eight-Minute Meditation," you can actually begin to cultivate mindfulness right now, by doing mundane exercises, and you can begin to do this in only one minute.

One minute. That is the starting point.

When new patients come to see me and know they are going to learn meditation, one of the biggest fears they have is how to find an hour a day to incorporate meditation into their already busy day. The success of meditation is not defined by the length of meditation but rather by the techniques of meditation. When I teach meditation to newcomers, I teach meditation in one minute, three minute, five minute and ten minute increments. Although there is value in lengthy meditations, especially when one has the luxury of time (often during the recovery process for surgery), most of what is truly valuable in meditation can come in much shorter periods of attention, practiced several times each day.

Let's begin to cultivate mindfulness.

A tip for doing these exercises it to purchase a digital timer, like that a cook may have in the kitchen. A decent digital timer can be purchased for less than ten dollars, and is a great tool for meditation.

EXERCISE 2:
ONE-MINUTE MINDFULNESS

Right now, as you read this book, take 60 seconds to pay attention in a particular way to your breath. The particular way that you are going to pay attention to your breath is nonjudgmentally, by simply experiencing what you experience as you experience it.

And so, sit on your chair, and focus on your breathing. In and out. Keep your eyes focused on a spot in front of you, or you can even close the eyes, whatever is most comfortable for you. And take a moment – 60 seconds – to pay attention to something that we rarely pay attention to, and that is the breath. Feel it. Feel it coming in through each nostril and out through the mouth. Notice the sensation of warmth or coolness in the breath, and what it feels like to pay attention to the breath. Continue to do this, and if any distractions, thoughts or feelings impede your focus on breath, simply recognize their presence and continue to breathe, recognizing them as neither important at this time nor unimportant.

You did it! You practiced cultivating mindfulness!

If you can do this exercise for one minute, then you can begin to do this for three minutes, then five minutes, and eventually ten minutes. But perhaps more important than earning your ten-minute diploma, you will also intuitively know how to stay focused on the present and nonjudgmentally see thoughts as just thoughts. This means the gravity of medical conditions can be just that, accepted in the moment as they are, even though the thought of dwelling on these realities can be overwhelming.

Acceptance is an important process in medicine and in the recovery process – acceptance of limitations, acceptance of unexpected outcomes and acceptance of serious consequences that may have even stemmed from earlier life choices. The book *Alcoholics Anonymous* again has a great thought on acceptance, and developing mindfulness through meditation is the method of achieving the acceptance that so many people struggle to achieve.

I explain to patients that acceptance does not imply liking a situation or scenario or outcome. Acceptance in no way implies that you want something negative to happen, but it does imply that you lay it out on the floor before you, see it as it is and move on from there with whatever you have at this moment. Here is what the book *Alcoholics Anonymous* says about acceptance:

"And acceptance is the answer to all my problems today. When I am disturbed, it is because I find some person, place, thing or situation - some fact of my life - unacceptable to me, and I can find no serenity until I accept that person, place, thing or situation as being exactly the way it is supposed to be at this moment.

Nothing, absolutely nothing happens in God's world by mistake. Until I could accept my alcoholism, I could not stay sober; unless I accept life completely on life's terms, I cannot be happy. I need to concentrate not so much on what needs to be changed in the world as on what needs to be changed in me and in my attitudes."

One can easily replace alcoholism and sobriety with other words. Try these sentences on and see if any of them fit:

- Until I could accept my cancer, I could not stay well.

- Until I could accept my surgery, I could not find healing.
- Until I could accept my pain, I could not feel security.

Until I could accept my limitations, I could not find freedom.

Mindfulness is a life-changing concept, one that does not take years to master, but rather a few intentional moments each and every day.

There are other ways to practice mindfulness that meditation instructors often teach. Since the idea is to pay attention to something in the present in a particular way, one could focus on the awareness of simple daily tasks such as eating or walking. One mindfulness meditation exercise given to new students of meditation is to walk from one side of the room to the other, simply being mindful of the process of walking, seeing the muscles of the body working together and being aware of something we do automatically in life with usually no thought to it. This is an exercise in development of the specific meditation skill we call mindfulness.

Chapter 5:
Concentration
Meditation

Concentration meditation is different than mindfulness meditation in that it focuses the mind on a specific thought or even a specific location, such as a repetitious mantra or a dot on the wall. Mindfulness is an exercise in awareness, but so is concentration meditation. In fact, concentration meditation is all about excluding awareness of other things, and bringing all of your being or attention to something neutral or mundane. Think of how wonderful mastering concentration meditation truly is. As a patient, I suffered relentless pain 24 hours a day, seven days a week, 365 days a year. Although the pain would be worse at some times than others, it was always present. Even though I applied every principle of mindfulness meditation, I was almost always aware of my pain. For those who suffer excruciating pain, you know how this permanent awareness of pain can devastate your spirit and ruin even the most precious moments of life.

I remember one experience in Prague with my son years ago. We were having a wonderful day, and found ourselves sitting on the famous St. Charles Bridge leading to Prague Castle. It was a wonderful sight in a wonderful city and I was sharing this with my son. As we sat to take a break in the center of the bridge we started humming. I think he started it first. Soon I joined into his tune though, and in a matter of minutes we found ourselves in Bohemia on the St. Charles Bridge, laughing and loudly singing the song "Bohemian Rhapsody." I even have a picture of us in this spot on that day. What a wonderful time, in Bohemia singing with my son, in one of the world's most historic places. But at that moment the pain from walking the uneven streets of Prague had exacerbated my condition, and no matter how wonderful the moment, that sensation of gnawing pain was with me, and we retired early to the hotel for some TV and web surfing because of that pain.

What concentration meditation does is it allows us to practice moving your attention from one thing to another. Back in my hotel, I spent some time doing concentration meditation, simply bringing my attention away from the pain and to a point on the drapes of my hotel. In fact, years later I can still picture in my mind the colors, spaces and texture of those drapes. And in bringing all of my attention and awareness to that spot, I was able to exclude my awareness of pain and begin anew. Over the next few days we continued to walk the streets, visited the shops selling gold and garnets and took a river tour. We ate in outdoor cafés and spent wonderful time together, and every time my awareness shifted to my pain, I would shift my awareness for a few minutes to a fixated point.

You can practice concentration meditation right now. Again, find a comfortable place free from distraction and dedicate the next minute to this simple exercise.

EXERCISE 3:
CONCENTRATION MEDITATION

As you relax in your chair pick a point on the far wall. It can be any point, a speck of paint, a color on the wall, a shadow or anything you become aware of. Now fixate your attention on that spot, not removing your eyes from it. You may notice you want to shift your gaze, and if you must for a moment you can, but quickly return your attention to that spot.

Now notice how easy it is to bring your attention to a specific point and to immerse yourself in this experience. See yourself being drawn to that point, and anytime an intrusive thought or awareness comes to mind, allow yourself to discard that thought in favor of your awareness of that spot. Experience that spot. Be a part of that spot. For just a few moments in time let nothing exist other than that spot. Now relax and congratulate yourself for learning a new meditation technique, concentration meditation.

It may seem too simple, but this is practice. If you intentionally practice the technique of directing your concentration to a specific spot, you will have the skills to automatically direct your attention to those things in life that are more pleasant. The reason this simple meditation works is that as you practice on a daily basis, the old habits of subconscious awareness of the same old things that you have noticed in life (i.e. pain), are replaced with intentional living and the ability to direct your thoughts.

When I was in Germany attending hypnosis training, I would fly to Germany from Tulsa on Tuesday, arrive on Wednesday, take Thursday off to adjust to the time change and then spend the entire weekend in class. I did this once a month. One day I arrived in Frankfurt, spent the night and then took my train to Speyer the next day and checked into my hotel. There was a dog outside my hotel window barking all night. I was desperately trying to flip my days and nights around so that I would not be jet-lagged in class, and this dog wasn't helping. He kept me up all night.

In the morning I came down to breakfast and into the classroom at the hotel where I was staying. The instructor knew me well, and said, "You must have just gotten in last night, you look exhausted!" I replied that no, I had gotten into Frankfurt on Wednesday and had plenty of time to rest, but that a barking dog had kept me up all night. I knew most of the class participants had stayed at the same hotel, so I looked to them to confirm my story, but they just had a bewildered look on their faces, and they all looked rested. The professor turned to me and said, "Well Richard, I guess you shouldn't have listened to him all night!"

Wow! What a lesson in shifting my awareness. The other guests stayed in the rooms next to me, with the same dog barking outside the same windows. But I was the one who chose to focus on him.

What concentration meditation can do for you is to stop the barking dog. Or more specifically, to teach you to be able to focus your concentration like a laser on something else other than what is bothering you. Remember though, pain is an indicator that something is wrong in the body. It tells us that a change needs to be made. Never overlook new, severe or difficult pain without being diagnosed by a physician.

CHAPTER 6:
RELAXATION

Relax! How many times have you been told to relax? Maybe just before the eye doctor directs a jolt of air into your eye? Perhaps just before a painful injection? Have you ever been in an argument and had someone tell you to just, "Take a breath and relax." That is of course, the most irritating time to hear this word. Maybe you have even told yourself, "Just relax, you can get though this..."

Relaxation is one of those things that is easier said than done. Our society does not naturally lend itself to relaxation. Computers were supposed to make thing easier, but now there are more things to break. iPhones and other smart mobile devices were supposed to make communication easier, but now everyone wants a response to a call, email, text message or Facebook post, NOW!

In 1929, a physician named Edmund Jacobson was able to prove the connection between excessive muscular tension and different disorders of body and psyche. He found out that tension and exertion were always accompanied by a shortening of the muscular fibers, that the reduction of the muscular tones decreased the activity of the central nervous system and that relaxation was the contrary of states of excitement and was well-suited for a general remedy and prophylaxis against psycho-somatic disorders. This first study in the medical benefits of relaxation gave birth to the methods of self-hypnosis and meditation that are being recognized as first-line interventions in medical treatment.

As you read this book, you are probably filled with high hopes for finding relief from the medical complications that have distressed you, and you probably have some happiness and excitement. But at the same time, the complications of the past, and just the stress of everyday life probably have you carrying tension in your body. If you are in acute pain, it may be easy to note the place in your body where you are carrying the tension of life. But even if you are not in acute pain, you can probably find a spot where you are carrying the tension of the day. Perhaps the back? Shoulders? Brow? We all carry tension physically and usually build it up in the same spot. We do this unaware of how much stress we are actually adding to our bodies each day.

Jacobson recognized two principles. The first is that people carry built-up tension that affects health. The second is that when people are taught the difference between tension and relaxation, they automatically begin to carry relaxation rather than tension. Essentially, the unhealthy but normal state is to carry tension automatically so that we have to make a conscious decision to relax. In learning what he called "Progressive Muscle Relaxation," the reverse becomes normal. In reversing this process, we begin to see tension as the intruder and automatically prefer to live in relaxation.

Another remarkable discovery he made is akin to the concept of "fractionation" in hypnosis. In hypnotherapy, when we create a trance state, then re-alert the patient, and quickly return him to a trance state, the trance becomes deeper. It is a lot like the alarm clock going off in the morning and hitting the snooze alarm. When the alarm first goes off, you usually feel awake and ready to go, but you know that you have ten more minutes. It is amazing how when that snooze alarm goes off ten minutes later, you often feel like you are more tired than when you first heard the alarm. So you hit the snooze button again. More often than not, when the alarm goes off for the third time, you drag yourself out of bed because you have to, more tired than when it first went off. Often people think to themselves, "I should have just gotten up and stayed up!" This is a demonstration of fractionation.

Likewise, when you tense and then intentionally relax a muscle, then tense the muscle again and relax it a second time, the relaxation experience doubles. Repeating this process in the various muscle groups is how we train our bodies to recognize relaxation and to develop deep levels of relaxation even when discomfort or stress have made it feel like it is impossible to relax.

EXERCISE 4:
RELAXATION MEDITATION

In this exercise we are going to practice recognizing the difference between tension and relaxation, and practice attaining a deep state of muscular relaxation. Sit comfortably in your chair, with the feet on the floor and the spine upright and away from the back of the chair. Scan your entire body for obvious tension and let that relax away. You can do this exercise with the eyes open or closed, as many prefer to practice with the eyes closed.

As your hands rest on your thighs, you are going to tense them into a fist and hold that tension. Not so tight that you feel pain, but tight enough to feel the fingers in the palm of the hand, and the muscles in the fingers, the back of the hand and wrist become tight. Now hold that tension, paying attention to what tension feels like and count to three. Then slowly release the tension, opening your fingers and letting them rest on your knees. As you do, notice the sensation of relaxation in the muscles, the tingle of relaxation and how it feels to let go of that tension.

Now repeat the process. Hold that tension in the fists, noticing and feeling the tension, counting to three and then relaxing. As you relax the second time, notice how deeply the muscles relax, almost twice as relaxed as the first time.

Congratulations, you have learned the basics of Progressive Muscle Relaxation, a skill that will serve you well in the management and recovery of medical conditions. And you have done what many people seem unable to do, since you have taken a moment for yourself and relaxed.

Doesn't it feel wonderful?

CHAPTER 7:
VISUALIZATION

Studies tell us that most people do well with visualization. For the majority of people, the visual learning style is the predominant learning style, in that people learn by seeing something. Visual imagery is a key resource for us in medical meditation because for the most part it is easy to do and something that you already know how to do, and because visual imagery activates the healing process in the body. How does it do this? Our minds are remarkable machines, capable of creating whatever we think. In fact, nothing exists today that was not first somebody's thought. At a metaphysical level, when one has the ability to visualize healing, the body has the ability to activate the hormones, enzymes and chemicals that promote healing.

Try a little experiment here as you read. In a moment relax your body and close the eyes. Imagine a fresh lemon when your eyes are closed. Create a vivid picture in your imagination of looking at that juicy, big yellow lemon and imagine cutting into it with a sharp knife. As the juices splatter as you break apart this plush, ripe lemon, imagine taking a bite into it as if it were an apple. With your mind imagine the taste, and feel the coolness of the lemon. Then open your eyes.

What happened to you? Did you pucker? Did you notice salivation? Did you move your head away from an imaginary lemon? Could you actually taste the sour?

I have done this experiment with hundreds of people, and the majority of them have some physical response to just the thought of this imaginary lemon. Thoughts do produce physical responses. As easy as it is to stimulate the saliva glands, it is easy to stimulate the body's healing potential.

If you are one of those few people who really struggles with visualization, know that by practicing visualization you can improve your skills in this area. It is something that can be learned.

I have listened to many guided meditation sessions where other people tell me what I should visualize. While these may be effective for some, the most effective visualizations come from within you. And so in the following exercise you may create the images I suggest in any way that you want to. This exercise helps move our time in meditation from one-minute mindfulness or concentration exercises to three-minute or even five-minute periods of meditation.

EXERCISE 5:
VISUALIZATION MEDITATION

In this exercise we are going to take a few minutes to practice visualization. Find a comfortable place to relax, scan your body for any tension and let those muscles go loose. Take in a deep breath or two, noticing how your heart rate slows and your breathing becomes smooth and rhythmic. When you are ready, close your eyes.

And with the eyes closed, imagine that you are outside under a clear blue sky. It is not too hot or too cold, and it can be a beautiful place you have been to before or a place you would like to go to, or simply a place of your own creation. Now take a moment to look into the sky and notice a single white puffy cloud lazily floating across the horizon. As it slowly moves across the sky it will become smaller and smaller, eventually drifting into the horizon and disappearing. Take another moment to be mindful of your experience with visualization and when ready, simply reopen the eyes, feeling refreshed and wonderful.

Go ahead, try the exercise outlined above. It is amazing how a simple break in the day to do a little daydreaming can actually impact our serenity. This is the skill of visualization. We can now use this skill to activate the healing potential deep within each one of us.

Up to this point I have taught you some very simple meditative techniques. If you simply review these exercises each day for the next week and never go any further, you will have some core techniques that can, when practiced and incorporated into life as automatic responses, have a profound impact on both your quality of life and your body's ability to heal. In the next section of this book I am going to build on these basic meditation techniques, and teach you some more advanced methods of medical meditation. In the last chapter I will create some routines to help you with preparation for surgery, and provide you with ideas following surgery that can benefit your recovery process.

It is so exciting to work with others and help them find health. I am glad that you have stayed with me so far and have tried the simple meditation exercises in the previous pages. Although mastering these techniques can take some time and practice, the benefits are almost immediate. Of course, the long term promises are faster recovery, fewer complications and reduced pain.

CHAPTER 8:
AUTOGENIC TRAINING

In this chapter, I am going to introduce you to a technique that I learned 20 years ago while working as a graduate student in an inpatient psychiatric unit. This unit was in a large medical and surgical hospital and the majority of the patients were elderly or had traditional illnesses and injuries that happened to be exacerbated by psychiatric difficulty. It was a great learning experience for me. I worked second shift, and the unit was run by old blue-haired nurses who had been nurses for many decades. In that type of environment, being moved to the psych floor was a reward for years of good service, since there is less heavy lifting involved. Many of the nurses were old school. They knew what worked and what didn't.

I was hired to do patient education, and assist the nurses in any way they needed my help. Each night at 9 p.m. I facilitated a relaxation training group, sometimes personally directing clients through various meditation processes, and other times playing a cassette tape. One of the blue-haired nurses named Mary gave me a tape called "autogenic training," and it soon became my favorite method of teaching patients meditation. To this day, it has been among the most helpful techniques that I have incorporated into my own meditation practice, and one of the tools that my patients derive the most benefit from.

Autogenic Training

At about the same time that Edmund Jacobson developed Progressive Muscle Relaxation, a German psy-chiatrist named Johannas Shultz published a book called "Autogenic Training." His methods preceded the development of biofeedback, and many of his original ideas have been incorporated into today's modern pain-control treatment programs. His method of Autogenic Training incorporated the meditative principles of visualization with his recognition of the mind's ability to influence its own autonomic nervous system. Shultz taught patients that *they* had control over how they felt and that through body awareness, exercises had the ability to change the way they felt.

Think about this: As a patient, do you ever feel helpless or out of control? Do you feel powerless over the outcomes of procedures and illness? Shultz taught a revolutionary concept - the concept of our innate ability to use our bodies to change perception.

For a short time in the early 1990's I worked as a family therapist in an eating disorders treatment program. Whether the diagnosis was bulimia, anorexia, bulrexia or another disorder, all of the patients had one thing in common: they felt that in life they had little or no control. Psychologists theorize that, at one level, eating is the only thing that such patients have control over and so for whatever reason, some patients engage in self-harming behavior, even behavior that can hasten death, in order to manifest control. Now there are many other factors that complicate and create eating disorders, but this prevalent psychological explanation rang true with me in my experiences on that unit.

I have met many other patients who feel powerless. Migraine patients have often been from doctor to doctor and test to test, taken a variety of medications and still find no relief. They are among some of the most powerless-feeling patients that I have met. Other patients I have worked with feel powerless in the dynamics and responsibilities of their families, the court system or job. Many sick people work simply to maintain health insurance benefits. Of course following my own surgery I felt powerless. I had to have my teenagers answer my phone, wheel me to my chair and let my dog in and out. Of my biggest frustrations was the effort to sit on the toilet, but if I stood up to urinate, holding my right leg out I always peed on my left sock at the end and left a small puddle on the floor for someone else to clean. Humiliation is a feeling that is closely related to powerlessness.

Autogenic Training is a confidence building meditation. It teaches that even the most powerless have ultimate control over what is most important - the body. Perhaps not at the level we hope for, but at some level, we always retain control, and autogenic training teaches this through experience.

In a few moments you will demonstrate to yourself one of the most remarkable abilities for self-healing by practicing autogenic training. *Auto* means "by oneself," and *genic* is a word that implies "from within." The changes that you experience here will be changes you create from within yourself, demonstrating the power of the mind to fully control our physical responses through the autonomic nervous system.

Take the next two or three minutes and guide yourself through this simple process.

EXERCISE 6:
AUTOGENIC TRAINING

Again, sit in your chair in the posture that promotes awareness and comfort, with your spine straight and your feet on the floor. This is a very brief exercise, based on the longer protocol for autogenic training. Begin by closing the eyes and focusing on the hands. In this exercise, it is okay to join the hands together by holding onto each other with the fingers, and sometimes people find when the hands are lightly joined that it makes this process easier. Now as you relax, focus on your hands and say to yourself, "My hands are warm and heavy. My hands are warm and heavy."

As you do this, say it out loud, focusing on the sensation of warmth in the hands and the sensation of heaviness. Allow yourself to feel warmth and heaviness as you repeat, "My hands are warm and heavy. My hands are warm and heavy." Now focus on your feet as your heart rate slows and your muscles relax. Say to yourself, "My feet are warm and heavy. My feet are warm and heavy." Let yourself concentrate on these sensations of relaxing as your feet feel warmth and heaviness. After a few moments experiencing the sensations of warm and heaviness, reorient to the room and open the eyes.

After completing this exercise, ask yourself, "Did I notice the change?" For some people the change in perception is very intense even the first time. For others the change is less intense, with only heat or heaviness predominant. That is alright, as Schultz laid out a complete protocol taking eight weeks to learn, but in just doing this one simple exercise, you have begun the process of learning that you control the responses of the autonomic nervous system.

For many people, the first time I guide them through a complete series of autogenic suggestions is empowering. They immediately feel a change and immediately recognize their own ability to control sensations of heat, heaviness, calm or coolness. What type of patient benefits from such a practice? Irritable Bowel Syndrome patients, pain management patients and patients in just about every medical setting.

A more profound question is this: if you can control sensations of coolness, warmth and heaviness, do you also have the ability to control pain? Or comfort? Or healing?

CHAPTER 9:
LIFESTYLE CHOICE IMPACTS

So far, we have practiced five core meditation skills:

1. Mindfulness
2. Concentration
3. Physical Relaxation
4. Visualization
5. Autogenic Responses

Each of these has value in and of itself. Even if you go no further in this book and practice just these principles, you will find them tremendously helpful to you. But recovery is wonderful, health is valuable and building on these skills with some more formal exercises will go a long way towards bringing about the life-changing benefits of medical meditation. Each of these exercises can be expanded on and each can be combined with others to achieve maximum benefit.

In addition to the principles of medical meditation, there are other lifestyle choices that you can make that will contribute to healing and health. Although most people are aware of these things, I want to point out that your goal in purchasing this book was to recover faster, to manage pain and to promote healing after surgery or during illness.

There are several things that one can also do to promote healing, and these include:

STOP SMOKING

Nothing impedes the body's natural ability to fight disease and healing than smoking. It is the number one contributing factor to slowing healing and experiencing medical complications. Even if you are having foot surgery, and you are young and your lungs to this point have been healthy, smoking will make the healing process slow. I use the meditation techniques in this book with many of my clients who see me for smoking cessation. These are the things I teach them to manage withdrawal, curb cravings and end the obsession with smoking. Everything in this book can be applied to recovery from drug addiction, including addiction to nicotine.

DIET

During periods of illness or surgery, we tend to be less mobile and utilize less energy. You can avoid adding weight by reducing your food intake during these periods. Following surgery, it is common to desire comfort foods, and allowing yourself such luxuries once in a while is perfectly okay, but you need to balance this with an awareness that your decreased physical activity requires less fuel (in the form of calories) to burn. Being overweight is another factor that complicates recovery, and adding on pounds during the recovery phase should be avoided.

EXERCISE

I am certified as a personal fitness trainer in addition to being a licensed psychotherapist. We can only function as well emotionally as we are physically, and for this reason exercise is important during the recovery phase. There isn't anyone who cannot do some form of exercise. In fact, preceding any surgery you should make yourself as strong as possible by participating in an exercise program. Following surgery, ask your doctor about what exercises you can do. Perhaps it will be simple arm or leg exercises, or moderate walking. But in general, getting out of bed and engaging in some physical activity at some level is good. My 97-year-old grandfather was lifting small weights in bed during the last year of his life, something that I am sure contributed to his comfort, health and happiness in his final year.

SUPPORT SYSTEMS

Having the right support following surgery is important. I was blessed with three wonderful teenagers, two with their own cars, a mother nearby, and a wonderful girlfriend who cared for me 24 hours a day. I know I drove them all crazy asking them to bring me the remote, help me into the tub and go on errands for my business. My recovery took place during the holidays, and each of them had their own responsibilities in life, but all of them, including my friend Steve, were available to help with a smile on their faces (a shout out to my lifelong friend Steve Lancaster, who spent all day Sunday doing household mechanics, changing light bulbs and hanging my son's guitar rack on the wall).

I meet many people who don't want to be a burden, and often fail to tell even close family or friends of impending changes or surgery, and often do not ask for help. These are the people who are seen in the emergency room later, who fell when walking without assistance, who didn't care for themselves as necessary and who complicate what can be a time of recovery with their pride.

Karma is a powerful law. People know this, and most are willing to help if we just ask them – if for no other reason because they know that one day they can count on you to repay the favor.

In addition to family and friends, support groups and community organizations can be of tremendous value. My office manger won her four-year battle with breast cancer. She would tell you that many things contributed to her victory, including self-hypnosis, medical meditation and restorative yoga. In fact, as I type this, she is at the surgeon's having new nipples created – the final procedure in what has been a long and painful four years. She remained an active participant in her church, but even as she faced life-and-death issues and derived tremendous help from the yoga classes offered by her hospital, she faced hostility and questions from those who did not understand her choice to utilize some of the techniques of medical meditation. Out of ignorance some even blasted her "occult" leanings, and as a result, she to some extent created new supports, preferring the understanding of other patients at the cancer center over those who to this point in her life had always been a traditional support – her church.

As a person both ordained in ministry and practicing as a mental health professional, I want to state that at no time do you have to defend the choices you make for yourself in order to do that which is right for your own health. If the supports that you expected to help you fail based on ignorance, a lack of understanding or even their own inability to face their own frailties, seek out new supports and in the end you will be free. As a patient, you only need your doctor's permission to utilize the proven methods of medical meditation, not your mother's or your minister's.

Chapter 10:
Extending Meditation
Practice

People who are new to meditation often seem overwhelmed by the prospect of sitting for 30 minutes "doing nothing." As you can see from the many exercises in the preceding chapter, the time spent in meditation is not "doing nothing," but rather using quiet stillness to learn and experience. Meditation is an active process. It is something you are engaged in. Using the exercises in this book, you will be engaged in mindfulness, concentration, relaxation, visualization or autogenic training exercises. Most likely you will be engaging in a combination of these things. Up to this point you have meditated for brief periods, in the first exercises for a minute or two, and in the later exercises for three to five minutes.

Although I do not think that the length of time spent meditating has any real impact on the efficacy of meditation, when meditating for longer periods of time one can actually practice more techniques, and I suppose that in the discipline of sitting for a period of time, learning can also take place.

What I am getting at here is that those who hold themselves out to be expert meditators just because of how long they meditated each day are really missing the point of meditation. I do occasionally set a timer when I do longer meditations, but generally when I begin the process of medi-tation, I do not know if I will be spending five minutes or forty-five minutes in the process. Rarely do I set time as my goal but instead it is the experience of meditation that I set as my goal.

You can experiment with the various exercises in the previous portion of this book and use a combination of them, or even do one technique for a longer period of time than suggested in the exercises to gain practice in doing longer meditation. For example, you may choose to practice mindfulness for three minutes or five minutes rather than the one minute in our first example. As you do this, notice how moments change and the feeling of each moment. Allow yourself to just see thoughts as thoughts, not worrying about having too many thoughts or an absence of conscious thought.

One of my favorite meditations combines concentration meditation with creative visualization. In the example I gave in the second exercise, I used the imagery of a single puffy cloud floating across the clear blue sky. You may choose any visualization that you find relaxing. Perhaps you want to create the image of a tropical paradise, or the majesty of a mountain sunrise. Feel free to edit my suggestions with your own ideas. In the following exercise, I combine concentration meditation with visualization, and then end with a positive affirmation.

Affirmations are positive tools for change and central to many methods of meditation and self-hypnosis. Affirmations do not need to be complex, and they can just affirm your progress or your state of serenity or your body's ability to heal.

EXERCISE 7:
EXTENDED VISUALIZATION

Begin by finding a comfortable place, such as sitting in a chair or lying on the floor or a bed. Give yourself permission to take a few minutes to yourself and to practice these new skills. Make sure the phone is off or that a sign on the door says do not disturb, and turn down the volume on your computer to avoid unexpected noises.

Now, as you sit comfortably, scan your body and let go of any obvious tension. Let any tension that remains drift out through the fingers and toes, and begin feeling a sense of calm. Notice that your breathing is already slower and your heart rate has begun to decrease.

Breathe deeply two or three times using the technique of deep breathing taught in Exercise 1, visualizing that balloon being filled with air and then exhaling as the air slowly is let out of the balloon.

Now that you are ready to enter a deeper state of meditation, choose a focal point on the far wall. Bring all of your attention to this spot or place, and be aware of how easy it is to shift your awareness to a small point on the far wall.

As your attention is focused on this point, let your concentration increase, simply letting any distracting thoughts remain in the background and continuing to focus on that point. Think for a moment of how it feels to set everything aside and focus on this point and notice any changes in the way you feel. As you continue to breathe, smoothly and rhythmically, you may notice that the eyes begin to water or tear, and when ready, just let them close.

At this point, begin your visualization experience, and imagine that you are outside under a clear blue sky. It is not too hot or too cold, and it can be a beautiful place that you have been to before or a place you would like to go to, or simply a place of your own creation. Now take a moment to look into the sky and notice a single white puffy cloud lazily floating across the horizon. As it slowly moves across the sky, it will become smaller and smaller, eventually drifting into the horizon and disappearing. At this point make a positive affirmation verbally and out loud, saying three times "I feel healing in my body. I feel healing in my body. I feel healing in my body."

45

Take another moment to be mindful of your experience with visualization and when ready, simply reopen the eyes, feeling refreshed and wonderful.

This is a simple method for combining modalities of meditation. You may adjust the combination and the exercises in any way that you find most beneficial. There is not a right way or a wrong way, there is only the way that meets your individual needs.

In this example, we have moved from brief meditation to 5- to 10-minute meditations depending on how you use your time. I now want to teach you a progressive muscle relaxation exercise that is lengthier than the one we practiced earlier, and also give you the full script for autogenic training. You can take a relatively short period of time to do the full PMR routine, or a very long time. It is entirely up to you. I personally tend to travel through the entire regimen of 16 muscle groups identified by Edmund Jacobson in about 12-16 minutes.

I will list the additional muscle groups that PMR is based upon, however please know that you do not have to do any groups that cause you pain or are difficult for you, and you can always adapt the methods in any way that helps you most.

Expanded Progressive Muscle Relaxation

Begin by sitting in the chair in a meditative posture, scanning the body for any obvious areas of tension. You will generally want to do this with the eyes closed, since it feels so good to practice with closed eyes, but in the beginning you will need to keep them open as you practice learning the various muscle groups.

There are many guided progressive muscle relaxation CD's available if you enjoy listening to the instructions at a pace set by a meditation guide. CD's are not necessary however, and you can do this with eyes open while reading from the page below, and then after practicing it several times and learning the various muscle groups, you can perform the exercise with your eyes closed.

RELAXATION SEQUENCE:

1. **Forehead.** Raise your eyebrows as high as they will go, as though you were surprised by something, and hold that tension while counting to three. Then relax those muscles, noticing the sensation of relaxation in the brow. Repeat.

2. **Eyes and cheeks.** Squeeze your eyes tight shut, not so tight that it is uncomfortable, but feeling the tension in the cheeks and eyes. Hold that tension and count to three, noticing what the tension feels like. Now relax, noticing what the sensation of relaxation feels like. Repeat.

3. **Mouth and jaw.** Open your mouth as wide as you can without experiencing pain in the jawbone, like you are yawning. Hold that tension and count to three, noticing what the tension feels like. Now relax, noting what relaxation feels like. Repeat. Notice how each time the exercise is repeated the sensation of relaxation doubles.

4. **Neck.** Be careful as you tense these muscles – do not strain or allow yourself to feel pain, just a light tension in the muscles. Do not do this if you have any history of neck injury or pain. Face forward and then pull your head back slowly, as though you are looking up to the ceiling. Hold this tension for a moment, and then relax, noticing what relaxation feels like. Repeat.

5. **Shoulders.** Tense the muscles in your shoulders as you bring your shoulders up towards your ears. Hold that tension, counting to three and noticing what tension feels like. Now relax, feeling the experience of relaxation. Repeat.

47

6. **Shoulder blades/Back.** Push your shoulder blades back, trying to almost touch them together, so that your chest is pushed forward. Hold this tension and count to three. Relax, taking a breath and noticing what relaxation feels like. Repeat. Again, notice how in repeating the exercise your awareness of relaxation is doubled.

7. **Hands and forearms.** Make a fist with your hands, hold that tension, and relax. Repeat.

8. **Upper arms.** Bring your forearms up to your shoulders to "make a muscle" and hold that tension in the biceps, feeling it in the triceps and all of the smaller muscles of the upper arms. Count to three and relax slowly, feeling the sensation of relaxation. Repeat.

9. **Chest and stomach.** Breathe in deeply, filling up your lungs and chest with air and holding that breath while you count to three. Now exhale, exhaling all of the air, feeling a sense of relaxation in the muscles of the chest and diaphragm. Repeat.

10. **Hips and buttocks.** As you squeeze your buttock muscles together, feel the tension in these large muscles, and hold that tension as you count to three. And relax, repeating this process.

11. **Upper legs.** Tighten your thighs. Hold tension – relax – repeat.

12. **Lower legs.** Do this slowly and carefully to avoid cramps. Pull your toes towards you to stretch the calf muscles. Hold tension – relax – repeat.

13. **Feet.** Curl your toes downwards. Hold tension – relax – repeat.

One can add affirmations following (or even during) any meditative process. In a previous example, a repetitive affirmation was made following visualization. After completing the program of tension and relaxation, affirmations can again be used. They can even be used as a brief meditative technique at any time during the day, apart from additional meditation methods.

An affirmation is a positive statement that you say to yourself several times each day. Many people who use affirmations make a list of 5-10 statements and write them on a card. They repeat the affirmations to themselves regularly. Incorporating these positive thoughts into your day is a way to reduce any anxiety that you may feel, because the positive thoughts replace the negative thoughts.

Here are some examples of positive affirmations. You may use these ideas, or write your own.

- I am calm and relaxed.
- I am free of all worry and stress.
- I can decrease my stress level.
- I take good care of my mind and body.
- I attract positive energy.

EXPANDED AUTOGENIC TRAINING

In an earlier chapter of this book, I described a brief meditation producing the sensation of warmth and heaviness in the arms and feet. This is referred to as autogenic training, and has its roots in yoga. Dr. Schultz described a complete program for mastering autogenic training in 1932, and there are many resources available on the web and in your public library for practicing the entire eight-week course. I have also created a training DVD guiding you through a full five-week course of autogenic training, and it is available at my website www.SubliminalScience.com.

Adding to the exercises given in the previous chapters, I want to outline a complete introductory session of autogenic training. It is designed to be practiced twice a day, for about ten minutes each time, for a period of one to two weeks. Successive programs add to this basic program, but mastering this basic program through repetitious practice will go a long way towards producing profound changes. I participate in an Internet hypnotherapy focus group. A thread was started by a therapist in the United Kingdom a couple of years back titled, "Benefits of a daily practice." In this thread he created a challenge that offered a prize to anyone who listened to his relaxation program each day for thirty days. He did this because it is so much easier to have people see for themselves the profound change that a daily practice brings rather than try to convince or explain the benefits to those who have not yet committed to a daily practice. This exercise is similar in that it doesn't seem profound, but it is. Practice this twice a day for two weeks or longer, and you will see results. And the results will be lifelong changes.

AN INITIAL PROGRAM FOR AUTOGENIC TRAINING:

Again, sit in a meditative posture. I prefer on a chair with the spine erect and the eyes closed. At first, you may need to keep the eyes open as you read these instructions, but soon you will commit them to memory and be able to close the eyes. This meditation consists of several phrases which you will repeat out loud three times. As you say each phase, allow yourself to experience the sensations described. A lot of people ask me what the solar plexus is. The solar plexus refers to what some call "the pit of the stomach." It is both a scientific term, specifically describing a nerve center in the body, and metaphysically referring to the central place where energy resides. In personal training, this is the "core" area where important muscles and nerves join together to work in unison to promote maximum functioning.

"My right arm is heavy and warm."
(Repeat three times)

"My left arm is heavy and warm."
(Repeat three times)

"My arms are heavy and warm."
(Repeat three times)

"My neck and shoulders are heavy."
(Repeat three times)

"My heartbeat is calm and regular."
(Repeat three times)

"My left leg is heavy and warm."
(Repeat three times)

"My right leg is heavy and warm."
(Repeat three times)

"My legs are heavy and warm."
(Repeat three times)

"My solar plexus is warm and comfortable."
(Repeat three times)

"My forehead is cool."   (Repeat three times)

"I am at peace." (Repeat three times)

As you expand the length of your meditations and begin to combine meditation methods, focusing on your breath becomes very important.

In one of the early exercises, I taught bell breathing, using the visual imagery of the helium-filled balloon. There are many techniques in yoga, *tai-chi* and meditation that can be used to help you develop either concentration on the breath or mindfulness on the breath. One of the simplest exercises is a way of expanding meditation focusing on breath work. It is a great breath-focused meditation because it is active.

As you recover from medical conditions or surgery, increasing your physical activity is very important. This exercise can help a person begin the process of becoming active, and with either a concentration or mindfulness focus, can expand awareness and promote health.

EXERCISE 8:
IGOR, AIRPLANE, UP

I call this exercise, "Igor, Airplane, Up." I am sure that someone else has come up with a more official name, but calling it "Igor, Airplane, Up" helps people learning it to remember the sequence of actions. In this exercise, begin by sitting in your chair in a meditative posture. You are going to do three motions with your arms, and draw in 1/3 of a breath with each motion.

Breathing through the nostrils if possible, extend your arms in front of you like the zombie character "Igor" in the old Boris Karloff movies. Then move your arms to the side like the wings of an airplane, drawing in another 1/3 of a breath. Bring your arms over your head, with the fingers pointing up to the sky, drawing in another 1/3 of a breath. Now as you slowly exhale through the mouth, bring your arms down to the side in a circular motion, resting them on the lap.

Repeat this two to three times, and you can take a moment between cycles to mindfully focus on your breathing and body, or to repeat affirmations that are helpful such as "My body is healing" or "I feel energy in my body" or "I am serene and content."

CHAPTER 11:
GETTING READY FOR
SURGERY

After my surgery, the biggest problem I experienced was feeling impatience coupled with frustration. I was frustrated by crutches, I couldn't get my walker through the narrow bathroom door, and when I told someone where to find something they didn't follow my directions and couldn't find it. I am normally a pretty easygoing guy, but I found my patience for others dwindling, and my feelings of helplessness added to my frustration. Having a house full of teenagers is not easy, but is even more difficult when you are off your feet for eight to 12 weeks.

I used the techniques of this book to control my emotions during the recovery period, and used the methods in this book to develop acceptance, a state of mind that I think is key to healing. When my manger was enduring her surgeries, she used these techniques to escape catastrophic self-limiting thoughts. Although she has now won her battle with breast cancer, a few years ago things did not look so good. The radiation that followed surgery and chemotherapy was wiping her out. I asked her what the prognosis was and I remember her telling me, "I stopped asking the doctors questions when I didn't really want the response and they quit giving me unpleasant answers." That is the heart of acceptance, and a key to recovery. She learned to live in the moment, mindfully and non-judgmentally living each day one day at a time. She acquired this mindset through meditation, learned by participation in yoga and through the practice of prayer consistent with her religion.

Prayer is one of the first meditative experiences that we are taught, and those who learn to pray in an affirmative way that draws on the supports that surround them do well. One of my best friends is an anesthesiologist and a fairly secular, or non-religious, person. He notes that all of his patients with religious faith, family and prayer do better managing anxiety, fear and emotions during difficult times. He attributes this to the meditative process of prayer, regardless of the patient's religious preference.

- I want to share with you the specific pathway to preparation for surgery, including the outcome-enhancing meditations I use with clients both pre- and post-surgery. I take a lot of referrals from people who have surgery scheduled in the next week or so, and so my time with them is limited. I usually teach them as much as I can in my office, and give them DVD's that expand on these methods with the instruction to view them and practice the methods following surgery.

- My recent foot surgery was a chance for me to practice what I tell my patients to do, and so I am going to reflect in the next few pages on my experience and teach you what I did that was so helpful. The surgery that I actually had was more complex than the one I expected. As I prepared for surgery, I knew that the more complex outcome was a possibility, so I prepared mentally for an extended period of recovery. After the surgery I used no painkillers, and began the healing process the second day home by listening to a guided meditation from a friend of mine.

- The first day home I was rather weak from the long surgery on the day before and using a guided meditation worked well for me. In fact, I listened to this particular guided meditation several times during that first week. Being stuck in bed, I found that meditation was one way to feel active and alleviate boredom. I also used the methods in the first few exercises of this book in one or three minute mini-meditations to reinforce the process of healing though meditation.

In practicing meditation both pre- and post-surgery with patients and in my own experiences, I find that the promises that the medical journals make regarding decreases in complications, a decreased need for medication and improved recovery are experienced by the majority of those I work with. In fact, my anecdotal experience is that 100% of those who actually practice the principles as I teach them have done well.

MEDITATION PRIOR TO SURGERY

As time is limited with many of my referrals, I try to spend at least two sessions helping them to prepare for surgery. If I have more time, I generally will work with someone on a weekly basis, teaching them as many techniques as possible and helping them to develop depth in their meditation practice. Chances are though that you have not anticipated your surgery (since many are non-elective), and picked up this book with little time to prepare and most likely little experience in meditation. Immediately preceding my foot surgery, I spent two days deviating from my usual meditative practices to return to these same basics that I teach patients who are new to medical meditation.

PRE-SURGICAL PREPARATION:
SESSION 1

In the first session with patients, I teach them the exercises that I shared in the first part of this book. These brief one- to three-minute meditations introduce the key methods and concepts of medical meditation.

Although we are not meeting in person, you have read through the book to this point and hopefully have practiced the various exercises. I will restate the exercises within these key areas, and if you have only read this book and have not yet practiced these fundamental exercises, please do so now.

EXERCISE 1:
BREATH WORK

Take a few moments now to sit on a chair in a quiet place. Allow yourself to be free from distractions for a few moments, turning off the phone or closing the door. Now scan your body, and anywhere you are holding tension let the tension melt away, relaxing the muscles in your body.

Now, take a few moments to practice deep breathing. Inhale through your nostrils as you envision that balloon being filled with helium, and then exhale by envisioning that air being let out. Repeat this several times.

EXERCISE 2:
ONE-MINUTE MINDFULNESS

Sit on your chair, focusing on your breathing. In and out. Keep your eyes focused on a spot in front of you, or you can even close the eyes, whatever is most comfortable for you. Take a moment – 60 seconds – to pay attention to something we rarely pay attention to, and that is the breath. Feel it. Feel it coming in through each nostril and out through the mouth.

Notice the sensation of warmth or coolness in the breath, and what it feels like to pay attention to breath. Continue to do this, and if any distractions, thoughts, or feelings impede your focus on breath simply recognize their presence and continue to breathe, recognizing them as neither important at this time nor unimportant.

EXERCISE 3:
CONCENTRATION MEDITATION

As you relax in your chair, pick a point on the far wall. It can be any point, a speck of paint, a color on the wall, a shadow or anything that you become aware of. Now fixate your attention on that spot, not removing your eyes from it. You may notice that you want to shift your gaze, and if you must for a moment you can, but quickly return your attention to that spot.

Now notice how easy it is to bring your attention to a specific point, and to immerse yourself in this experience. See yourself being drawn to that point, and anytime an intrusive thought or awareness comes to mind, allow yourself to discard that thought in favor of your awareness of that spot. Experience that spot. Be a part of that spot. For just a few moments in time let nothing exist other than that spot. Now, relax and congratulate yourself for learning a new meditation technique, concentration meditation.

EXERCISE 4:
VISUALIZATION MEDITATION

In this exercise we are going to take a few minutes to practice visualization. Find a comfortable place to relax, scan your body for any tension and let those muscles go loose. Take in a deep breath or two, noticing how your heart rate slows and your breathing becomes smooth and rhythmic.

When you are ready, close your eyes. With the eyes closed, imagine that you are outside under a clear blue sky. It is not too hot or too cold, and it can be a beautiful place you have been to before or a place you would like to go to, or simply a place of your own creation.

Now take a moment to look into the sky and notice a single white puffy cloud lazily floating across the horizon. As it slowly moves across the sky, it will become smaller and smaller, eventually drifting into the horizon and disappearing.

Take another moment to be mindful of your experience with visualization and when ready, simply reopen the eyes, feeling refreshed and wonderful.

EXERCISE 5:
RELAXATION TRAINING

In this exercise, we are going to practice recognizing the difference between tension and relaxation, and attaining a deep state of muscular relaxation. Sit comfortably in your chair, with the feet on the floor and the spine upright and away from the back of the chair. Scan your entire body for obvious tension and let that relax away. You can do this exercise with the eyes open or closed.

As your hands rest on your thighs, you are going to tense them into a fist and hold that tension. Not so tight that you feel pain, but tight enough to feel the fingers in the palm of the hand, and the muscles in the fingers, the back of the hand and wrist become tight.

Now hold that tension, paying attention to what tension feels like, and count to three. Then slowly release the tension, opening your fingers and letting them rest on your knees. As you do, notice the sensation of relaxation in the muscles, the tingle of relaxation and how it feels to let go of that tension.

Now repeat the process. Hold that tension in the fists, noticing and feeling the tension, counting to three and then relaxing. As you relax the second time, notice how deeply the muscles relax, almost twice as relaxed as the first time.

EXERCISE 6:
AUTOGENIC TRAINING

Again, sit in your chair in the posture that promotes awareness and comfort, with your spine straight and your feet on the floor. This is a very brief exercise, based on the longer protocol for autogenic training. Begin by closing the eyes and focusing on the hands.

It is okay to join your hands together holding onto each other with the fingers, as sometimes people find that when the hands are lightly joined it makes this process easier.

Now as you relax, focus on your hands and say to yourself, "My hands are warm and heavy. My hands are warm and heavy." As you do this, say it out loud, focusing on the sensation of warmth in the hands and the sensation of heaviness. Allow yourself to feel warmth and heaviness as you repeat, "My hands are warm and heavy. My hands are warm and heavy."

Now focus on your feet as your heart rate slows and your muscles relax. Say to yourself, "My feet are warm and heavy. My feet are warm and heavy." Let yourself concentrate on these sensations, relaxing as your feet feel warmth and heaviness.

After a few moments of experiencing the sensations of warm and heaviness, reorient to the room and open the eyes.

PRE-SURGICAL PREPARATION:
SESSION 2

In this session, I want to help you use the techniques you have found most valuable and to begin to see how when preparing for surgery, these techniques can be adapted to specific tasks related to surgery. Ask yourself which of the techniques practiced since session 1 were most valuable to you.

In private sessions, this is my starting point for tailoring individual programs to meet specific needs. I then provide additional instructions in session-combining methods, or expand on autogenic training or progressive muscle relaxation. I stress that these techniques are to be practiced, at least twice a day every day until the day of surgery. I also begin to adapt some of the imagery and ideas to unique medical situations.

I am sure you have already thought while reading the pages of this book of the many ways you can adapt these ideas to your specific situation. As a meditative tool, visualization can be used to create a mental vacation or safe place that one can emotionally or mentally retreat to during stress. But the same technique of visualization can be practiced and applied to specific aspects of healing and wellness. These can be very powerful.

For example, as you anticipate your surgery, do you see catastrophic images in your mind? In your anxiety, do you see the doctor operating on the wrong part, or do you create mental pictures of medical professionals scrambling to avoid medical emergency? These thoughts are not strange or bizarre, and I think most people actually have them. One who has practiced visualization can use the technique to create a mental safe-place or stress-free environment, but you can also use the ability to direct your visual imagery and "change the channel" on these distressing thoughts.

Changing the channel is important. As I mentioned earlier, all realities were once thoughts. It is important for us to maintain healthy thoughts. One must direct visualization of thoughts to images that promote optimal health.

EXERCISE 9:
HOW TO CHANGE YOUR MENTAL CHANNEL

This exercise uses a combination of concentration meditation with visualization to alter the images that sometimes cause distress and to promote imagery consistent with positive surgical outcomes.

Begin in the meditative posture of sitting in a chair with the head, shoulders and back aligned. Scan the body for obvious tensions, releasing those stresses. Find a place to focus your attention on the far wall, and bring all of your energy, thought and attention to that spot. As your breathing slows and your heart rate becomes more smooth and rhythmic, allow yourself to feel a sense of serenity and peace. After a minute or so, let your eyes close and create in your mind the following visualizations, almost as if you are watching a movie of you on that spot where you were focusing.

Imagine yourself the morning of surgery, feeling positive about definitive outcomes and resolution to pain, injury or illness. See yourself making the decision to proceed with surgery and being calm and ready. Whatever we think, we can create, and so make these images as real in your mind as possible, only thinking of positive feelings and actions.

Imagine yourself on the hospital bed, prepared for surgery in your gown and speaking to the anesthesiologist. Your surgeon and your anesthesiologist will both talk with you before surgery, and project this future image of yourself talking with them and being reassured by them and being guided by the anesthesiologist into a deep and peaceful state. Imagine yourself comfortably drifting into a natural trance state or meditative state guided by the anesthesiologist, knowing that you are in the hands of a well-trained physician and that you can choose positive thoughts and awareness as you drift into a state of unconsciousness.

Say to yourself, "I can choose the thoughts I take into surgery with me, and I will choose positive thoughts." Now visualize yourself asleep, while your skilled surgeon does what he was trained to do. See the doctors and nurses acting as the professionals they are, and if at anytime distressing images come to mind, recognize that they are distortions of your ideals and push them aside, refocusing on that spot where positive images are created.

In your mind's eye, imagine being wheeled to the recovery area where family or friends may greet you as your nurse and anesthesiologist monitor your recovery. As you imagine this scene, see yourself becoming revived and energetic and feeling supported by the people who are with you. Allow yourself to simply create this image of the surgical process and the excellent care you will receive, emerging from the process fully alert and oriented, rested and at peace with the surgery you have experienced.

This exercise is a powerful adaptation of the earlier techniques in this book because it creates a new reality for you. Instead of a reality with anxiety of the unknown, it is a metaphysical principle that if you can conceive in your mind a positive outcome, that reality follows these thoughts and the process of your surgery will be positive.

Chapter 12:
Using Meditation Post-Surgery

I am a believer in affirmation and positive thought. We create with our thoughts, and positive thoughts create favorable scenarios. Although you may have loss following surgery or outcomes that are not as expected, a focus on the positive can bring acceptance and joy even in the midst of turmoil and difficulty.

We can think ourselves sick, and actions at a subconscious level are often consistent with thoughts. For this reason, positive energies and positive thoughts are essential during the first few days of recovery. I cannot explain it, but thoughts impact outcomes. The body can at some level think itself well as easily as it can think itself sick. You can actually look at the glass being half full or half empty, and although it may be difficult or perhaps not your usual nature, the practice of meditation can make this paradigm shift.

Immediately following surgery, guided meditations using CD's or mp3's can be helpful. Sometimes the energy to think or create is hard to muster in these first few days.

After a few days, meditation can be used specifically to help you with several things. Outcome studies show decreased complications, faster recovery and less reliance on pain medication. I think one can focus meditations specifically in these areas.

EXERCISE 10:
MEDITATION TO PROMOTE SELF-ACCEPTANCE

In this exercise, begin in the meditative posture and relax. Use the fundamental outline of progressive muscle relaxation to guide yourself into a deep state of relaxation. If you have been practicing the difference between tension and relaxation over the past week or so, it should be easy for you to notice relaxation and to allow yourself to relax the various muscles, even without tensing any muscles. It is not important that you remember the entire protocol for all of the muscle groups, so just begin by relaxing the muscles in the brow and face, and extend this relaxation into the neck and shoulders, feeling relaxation travel through the hands. Extend that sense of relaxation into the back and legs, and to the little muscles of the feet. Sense now the experience of physical relaxation, and say positive affirmations to yourself. You may create any affirmation that would be beneficial to you. For example:

> "Like the lilies of the field, I am cared for by God."

> "I am a good person."

> "I am serene and secure."

Generally when I use affirmations, I create a series of three affirmations, and repeat each three times, paying attention to feeling the affirmation as much as the words of the affirmation.

You might not even believe the affirmations with the emotional part of the mind, especially when blame surrounds accidents or illnesses. It is not necessary to emotionally believe the content of the messages. I had an old friend who often said, "Fake it 'til ya make it," and this is a good time to be reminded of this phrase. As we reprogram towards positive self-talk, it will become part of both our intellectual and emotional belief system with repetition.

EXERCISE 11:
MEDITATION TO PROMOTE HEALING WITH AUTOGENIC TRAINING

In this exercise, you are going to take the suggestions from autogenic training and replace cool, warm and calm with suggestions directly related to recovery and healing. It is just as easy to feel a sensation of warmth or coolness as it is a sensation of decreased swelling, or of bone fusion, or of being able to urinate or of any other sensory experience that promotes healing. *Examples (repeat each line 3 times):*

"My right arm is light and free from swelling."

"My left arm is light and free from swelling."

"My arms are light and free from swelling."

"My neck and shoulders are strong."

"My heart is bringing oxygen to every cell in my body."

"My left leg is healing and strong."

"My right leg is healing and strong."

"My legs are healing and strong."

"My solar plexus is supplying my body with positive life force."

"My wound is closing and healing."

"I am at peace."

Any of these sensations can be developed using autogenic training, and you can adapt to your own situation any suggestions that promote physical healing as in the examples listed above.

EXERCISE 12:
MEDITATION TO REDUCE
PAIN WITH MINDFULNESS

Every time I go to the doctor or see a nurse, they ask me to rate my level of pain from one to 10. They actually use a pain assessment scale with frowning and smiley faces called the "Wong-Baker FACES Pain Rating Scale." If you have had chronic pain or multiple surgeries, or visited a doctor recently with any complaints, you have probably seen this. As a hypnotherapist who believes that professionals should be very careful about the unintended suggestions we give, I wonder why they didn't create a comfort scale. Why do they heighten my awareness of pain in the assessment and intake process? "Richard, how much pain are you in today?" they ask, often without realizing that the very question forces me to be aware of my pain and that this in and of itself will raise my pain level. I don't understand why the medical community is not trained to ask the question, "Richard, how comfortable are you today?" That question, even if I were in pain, would force me to recognize that at some level I am probably also comfortable.

Shifting our awareness is a process of mindfulness meditation. You can be both in pain *and* comfortable. You can be both depressed *and* happy. You can have grief *and* acceptance. These things are *not* mutually exclusive, and mindfulness meditation is a great tool for reducing pain by increasing our awareness of comfort.

Try this:

As you sit in a meditative position, be mindful of the present. In the past you may have judged your pain on various scales or been trained to recognize its intensity. Now is not the time to do so, now is the time to just be. If you say to yourself "I feel pain," nonjudgementally accept that as it is at this moment without attaching meaning to it. Notice that at the same time you may feel pain, you also feel other things. What are these other things? Do you feel comfort? Do you feel other emotions? Put aside yesterday, and focus on the now. At this very moment, just be, and mindfully pay attention to this being as it is.

The exercise is simple and profound. For those who really practice the methods, you will find that acceptance is the result.

CHAPTER 13:
IN CLOSING

This book has been a collection of ideas and exercises. The material presented here has helped countless patients of mine in a variety of settings, including my friends and myself. My hope for you is that you will practice the principles contained in this book and benefit from them. There is a lot of information here, and developing a regular meditation practice following recovery will do two things for you:

It will promote continued wellness and health.

It will prevent disease, discomfort and difficulty.

When I shared a rough draft of this book with my office manager, her comment was, "I wish I had this years ago. I want to share this with my restorative yoga instructor!" People are excited about meditation and the medical benefits of meditation. This book is a starting point, and my sincere hope is that you go on from here to experience wonderful things in life, one day at a time.

~ Richard Nongard

RESOURCES FOR MORE INFORMATION

For those seeking professional training at an advanced level in medical meditation and stress management consulting, please visit www.MedicalMeditation.org

Richard Nongard, Ph.D. is also a popular educator in clinical hypnosis and hypnotherapy. His books, mp3s and training programs are available for both professional and individual benefit, and may be found at www.SubliminalScience.com

Licensed mental health professionals seeking state board-approved continuing education in topics related to professional ethics, hypnotherapy, medical meditation and other mental health related issues can visit www.FastCEUs.com

KEYNOTE SPEAKING AND STAFF TRAINING

Is your organization looking for a dynamic and positive presenter? Richard Nongard has presented to some of the largest conferences and conventions in the mental health field, and to corporate wellness centers and community groups both in the USA and around the world. If your need is for a short presentation to motivate and empower any group, he is the ideal fit for your conference or convention. If you need corporate stress management training or host in-service seminars, you will find his lighthearted and practical style refreshing and well received. Call 1-800-390-9536 for more information.

DOWNLOAD THESE EXERCISES

All 12 Exercises in this book plus the Expanded Progressive Muscle Relaxation and Autogenic Training Sessions are available for audio download via Instant-Access mp3 files.

If you would like to experience these exercises in the form of guided meditation lead by Richard Nongard himself, please visit

www.SubliminalScience.com

Made in the USA
San Bernardino, CA
20 September 2017